Alessandra de Souza Conceição
Iranara costa da S.Teixeira

Popular uses and industrial applications of castor oil

Alessandra de Souza Conceição
Iranara costa da S.Teixeira

Popular uses and industrial applications of castor oil

A review of the literature

ScienciaScripts

Imprint

Any brand names and product names mentioned in this book are subject to trademark, brand or patent protection and are trademarks or registered trademarks of their respective holders. The use of brand names, product names, common names, trade names, product descriptions etc. even without a particular marking in this work is in no way to be construed to mean that such names may be regarded as unrestricted in respect of trademark and brand protection legislation and could thus be used by anyone.

Cover image: www.ingimage.com

This book is a translation from the original published under ISBN 978-613-9-77654-2.

Publisher:
Sciencia Scripts
is a trademark of
Dodo Books Indian Ocean Ltd. and OmniScriptum S.R.L publishing group

120 High Road, East Finchley, London, N2 9ED, United Kingdom
Str. Armeneasca 28/1, office 1, Chisinau MD-2012, Republic of Moldova, Europe
Printed at: see last page
ISBN: 978-620-6-59667-7

We dedicate the following work to our parents, who have always supported us, giving us the encouragement we needed to carry on even in difficult times of discouragement and tiredness, in short, for everything they represent in our lives.

Thank you

- We thank God first and foremost for blessing our journey, not only in these years as university students, but also in all the challenges we've had to overcome - he is undoubtedly the greatest teacher you can have;

- To the Adventist University Centre of São Paulo (Unasp);

- To our advisor Prof Dr Marco Aurélio Sivero Mayworm;

- To all the friends who helped us by giving us strength, words of comfort in times of difficulty, in short, to all those who directly or indirectly played a part in our formation, thank you very much.

Summary

The castor bean tree is a plant that produces oilseeds and is of Asian origin. Its ease of adaptation has allowed it to be cultivated in Brazil, especially in semi-arid regions.

This paper aims to present a literature review on castor oil (*Ricinus communis* L.) and its industrial applications as well as its use in traditional and modern medicine. To this end, a bibliographic survey was carried out of material published between 2002 and 2017, using databases such as Google Scholar and the Proquest platform.

The material surveyed presented various aspects relating to the use of *Ricinus communis* L. oil and its properties. As explored in the course of this work, castor oil has been used in a variety of activities such as biofuel production, biotechnology, medicine and ethnobotany, among others. Using this raw material, it is possible to generate many products, which is why the oil taken from the castor bean known as castor oil reaches a significant range of industries, from relatively simple products to sophisticated ones. This work contains information on how castor oil can be used in a versatile way, for example, in the treatment of stomatitis, in the disinfection of dental prostheses, for anti-inflammatory purposes, in the thermal insulation or waterproofing of roofs and buildings, as well as being used as a biofuel. These are just a few of the applications of castor oil that we have reported on and even though we have carried out a review of the most up-to-date literature, there is still much to be discovered and reported on this important species.

Key words: *Ricinus communis, Biofuel, biotechnology, medicine, industry, Ethnobotany, Toxicity, castor oil, castor bean, cosmetology.*

CHAPTER 1

Introduction

The castor bean tree, scientifically known as *Ricinus communis* L., is a plant from the Euphorbiaceae family and its fruit is known as castor bean or castor, and can have variable characteristics as shown in Figures 1, 2 and 3. In Brazil, it is also known as carrapateira, bafureira, baga, planta "christi" and palma-criste. In some African regions, it is known as abelmeluco. In English, it is called "castor bean" and "castor seed" (EMBRAPA, 2012 *apud* Fereira 2013). The castor bean tree is a species traditionally cultivated in the semi-arid regions of Brazil. It has economic and social importance and many uses in industry. It is a species of Asian origin, but can be found in various regions of Brazil, from Amazonas to Rio Grande do Sul (Beltrão and Oliveira, 2008).

Figure 1 - Green papaya fruit with thorns. **Source**: Máira Milani
Figure 2 - Castor bean fruit without thorns. **Source**: Máira Milani

Figure 3 - Pink papaya fruit with thorns.
Source: Máira Milani

Castor beans have been used in our country since the colonial era, when they were used to lubricate sugar cane mills. Its extraordinary adaptability, the multiplicity of industrial applications for its oil, the value of its cake as a fertiliser and protein supplement and the trend towards rising prices have made castor bean one of the most important oilseeds today (Schneider, 2003).

Among the various oil seeds, the castor bean has advantages due to its low production and implementation costs, favouring its production by small farmers in locations such as the north-eastern region of Brazil. Its main product is castor oil, extracted from its seeds, shown in figure 4, which is widely used in the cosmetics industry and in the automotive industry as a lubricant for high-speed engines (Oliveira Filho, 2013).

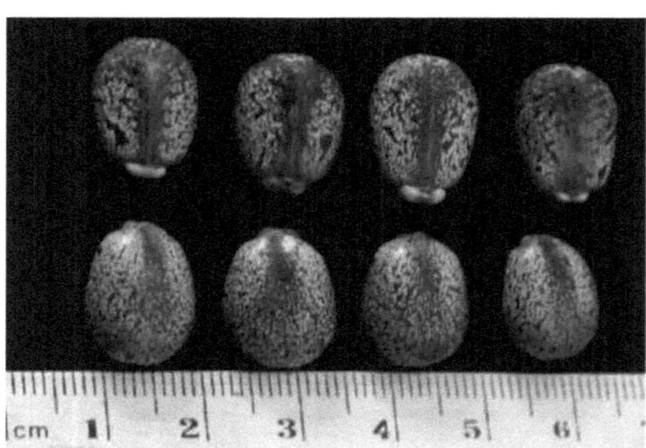

Figure 4 - Castor bean seeds
Source: Máira Milani

Plants of the species show great variability in various characteristics, such as growth habit, stem colour, size, colour and oil content of the seeds and leaf colour, as shown in Figures 5 and 6.

You can therefore find botanical types with a low or arboreal size, an annual or semi-perennial cycle, with green, red or pink leaves and stems, with the presence or absence of wax on the stem, with fruits with or without thorns, dehiscent or indehiscent, with seeds of different sizes and colours and different oil contents.

It has a fistulous root system, consisting of a pivoting main root, whose development varies with the size of the cultivar

Figure 5 - Purple leaves of the papaya tree, without wax, narrow leaflets.
Figure 6 - Papaya leaves, green with wax, narrow leaflets.
Source: Máira Milani

The secondary roots are well developed, but in dwarf plants they are more branched and penetrate deeply into the soil. The papaya tree is considered a drought-tolerant plant, but it needs around 500 mm of rainfall to achieve good yields.

Before flowering, it requires around 100 mm per month, distributed regularly over the first four months of the cycle, so that the bunches bloom under conditions of water availability.

The stem is thick and branched, ending with a raceme-like inflorescence. The main, or primary, stem, which may or may not be covered in wax as shown in figures 7 and 8, grows vertically, without any branching, until the first inflorescence appears, called the main bunch.

The lateral branches develop from the axil of the last leaf, just below the inflorescence. Wax is more abundant in young plants and those under water-stressed conditions (Milani, 2017).

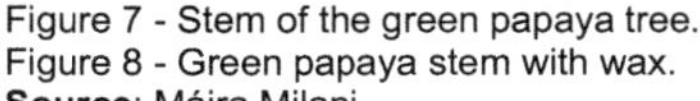

Figure 7 - Stem of the green papaya tree.
Figure 8 - Green papaya stem with wax.
Source: Máira Milani

In Brazil, castor beans are basically grown in two systems. In the traditional system, the crop is harvested by hand using medium and tall cultivars, as shown in Figures 9 and 10. This system is very widespread among small producers, but production can also take place using modern technologies in which large areas are cultivated and harvesting is mechanised (Moro, 2008).

Figure 9 - Low-growing papaya tree.
Figure 10 - Tall papaya tree.
Source: Máira Milani

"Under rainfed conditions, castor beans produce 1,200kg/ha of berries, with an oil content of 47%. Castor bean is a crop with great social appeal, because as well as producing oil, it can be intercropped with other crops such as beans, peanuts, cowpeas or maize" (Da Silva, 2017).

As well as being used in the chemical industry and to reduce dependence on oil, castor beans can be used to supplement the income of small and medium-sized producers.

These characteristics highlight the importance of this oilseed, the correct exploitation of which can have a positive impact on society, the environment and industry.

Castor oil is a raw material with unique applications in the chemical industry due to the peculiar characteristics of its molecule, which make it the only naturally hydroxylated vegetable oil with a composition predominantly made up of a single fatty acid, ricinoleic, giving it atypical chemical properties such as high viscosity and solubility in alcohol at low temperature (BELTRÃO, 2004 *apud* Fereira et al. 2013). Given this diversity of uses, the aim of this study was to review the literature in search of studies showing the possible uses of *Ricinus communis* L. oil.

CHAPTER 2

Methodology

In the course of this work, various pieces of information are gathered on the use of *Ricinus communis* in the following areas: biodiesel production, biotechnological products, use in ethnobotany and medicinal use, so as to be of great help in carrying out future work on the oil extracted from this species full of unique characteristics.

This work is a literature review based on a consultation of online articles in the following databases: Google Scholar and the Proquest platform, respectively. The searches were carried out using the following keywords: *Ricinus communis* biofuels, biotechnology, medicinal use and ethnobotany.

The articles used in this review were selected on the basis of the following selection criteria: objectivity, relevance of the results obtained in the research and the possibility of comparing the results of the article with related work. After selection, the articles were separated by theme and sub-theme, and the information gathered from them was organised in a logical and coherent manner, enabling a well-founded discussion and the creation of a clear and informative review.

The articles selected cover a period of 15 years and were published between 2002 and 2017, using a total of 49 bibliographical references that provide information on *Ricinus communis* L. and its applications.

CHAPTER 3

Development

3.1. The use of castor oil in biodiesel production

The main product of the castor bean is the oil extracted from its seeds, which is considered a noble oil with great industrial versatility. This is largely due to the fact that it contains a large amount of ricinoleic acid, on average around 90 per cent of its composition, with three highly reactive groups that together enable specific qualities for the production of a multitude of industrial products (Torres, 2006 *apud* Almeida et al. 2015).

The possibility of producing biodiesel from castor oil has created a new market for this product capable of absorbing a large part of the production of countries that have used this raw material, such as Brazil (Rizzi et al. 2010).

Renewable energy sources can and should be used sustainably, i.e. in a way that has minimal impact on the environment (Pacheco, 2006 *apud* Ferreira et al. 2013).

Castor oil is considered to be one of the most versatile from an agricultural and industrial point of view, with a usefulness only comparable to that of petrol, with the advantage of being a renewable and cheap product. Today, ricinochemicals are responsible for the production of more than four hundred products derived from this oil. Its greatest importance, however, lies in its potential as a raw material for biodiesel production and as an energy source (Góes, 2006).

The need to look for ecological alternatives for energy production has led to vegetable oils gaining prominence as a way of producing fuel.

> "Biodiesel is a biodegradable fuel derived from renewable sources, which can be obtained by different processes such as cracking, esterification or transesterification (...). This process also produces glycerine, which is used to make soaps and various other cosmetics. There are dozens of plant species in Brazil from which biodiesel can be produced, such as castor beans, palm oil, sunflower, babassu, peanuts, jatropha and soya, among others" (Ministério de Minas e Energia, 2017).

Biodiesel is an oxygenated fuel, which is why it has a very complete burn that allows for a reduction in pollutant emissions from the engine. When there is a biodiesel/diesel mixture, the higher the percentage of biodiesel present in the mixture, the lower the reduction in pollutant emissions (Conceição et al. 2007 *apud* Oliveira 2013).

In the biodiesel production process, the main product is an ester (biodiesel) and the main by-product is glycerine, which is used by the plastics, lubricants, cosmetics, pharmaceuticals and explosives industries. According to the ANP (National Agency of Petroleum, Natural Gas and Biofuels) statistical yearbook, 2016, the production of glycerine derived from biofuel in Brazil totalled $346,839m^3$ in 2015.

From a technical point of view, oil extraction and biodiesel production can be carried out in small plants using presses like the one shown in figure 11. However, this alternative is only viable in special situations, as the cost of production is much higher than in a larger plant (Embrapa, 2017).

Figure 11 - Press for extracting castor oil in mini mills.
Source: Embrapa

The main aspects to consider when deciding to install a mini biodiesel plant, such as the one shown in figure 12, are the high electricity and thermal energy costs of running the machines, the need to refine the oil and the control of the fuel's quality, which is done through very expensive laboratory analyses (Embrapa, 2017).

Figure 12 - Mini biodiesel plant for small-scale production.
Source: Embrapa

As far as energy generation is concerned, castor bean oil is well suited to the National Biodiesel Programme which, in addition to social inclusion, aims to reduce both Brazil's imports of petrodiesel and the burning of fossil fuels (to avoid, respectively, the evasion of foreign currency and the heating of the earth). In this sense, it should be noted that in addition to the socio-economic advantages, this oilseed has a higher oil content than the others and according to (EMBRAPA, 2004 *apud* Oliveira Filho 2013), each hectare cultivated with castor bean absorbs ten tonnes of carbon dioxide, which is four times the average for other oilseeds.

With this ability to absorb more carbon dioxide than other oilseeds, the castor bean tree is able to accumulate a lot of CO_2 (carbon dioxide) and becomes a strong competitor for biodiesel feedstock.

Miranda (2011) compared three raw materials for biodiesel production: Soya, Jatropha and Castor bean. The data from the energy analysis shows that soya oil is the most viable in terms of energy and the author recommends that to make castor oil more competitive in terms of energy, investment should be made in improving the separation processes. The author also states that Jatropha and Castor Oil cannot be disregarded as raw materials because they do not compete with food commodities and are highly productive in Brazil.

According to the Monthly Biodiesel Bulletin published by the ANP in 2016, the national production of biodiesel in our country was approximately 335,453 m^3 , considering the months of January to August, with the North region being the smallest producer and the

13

Centre West region the largest, with an average monthly production of approximately 152,939 m3.

Melo et al. (2008) argue that the extraction of oil from castor bean seeds for biodiesel production is responsible for the daily generation of tonnes of solid waste formed by the pressed seeds, known as castor bean cake. This waste is high in protein and is produced in the proportion of 1.2 tonnes for every tonne of oil obtained, which corresponds to 55% of the average weight of the seeds.

They state that when castor bean cake undergoes a hydrolytic process under acidic conditions, it can be used to produce ethanol, as it produces around 25.3g of ethanol per 100g of castor bean cake and suffers a side effect that makes it free of lethality when tested on mice. Based on this data, castor bean cake can be transformed from a waste product into a raw material for ethanol and feed.

These results indicate the possibility of developing an industrial model in which the stages of extraction, ethyl esterification with ethanol produced from castor bean cake and chemical saccharification of castor bean cake for feed production are connected.

The process of producing biodiesel from castor beans, like any other raw material, consists of the following stages: preparation of the raw material, transesterification reaction, phase separation, recovery and dehydration of the alcohol, distillation of the glycerine and purification of the biodiesel.

When biodiesel is produced from castor beans, the first step in preparing the raw material is not the neutralisation and drying process, when that is necessary. It starts much earlier, in other words, choosing the area for planting the castor bean tree, preparing the soil, planting the castor bean tree, cultivation, intercropping, production and harvesting the berries (Góes, 2006).

3.2. Process for producing biodiesel from castor beans

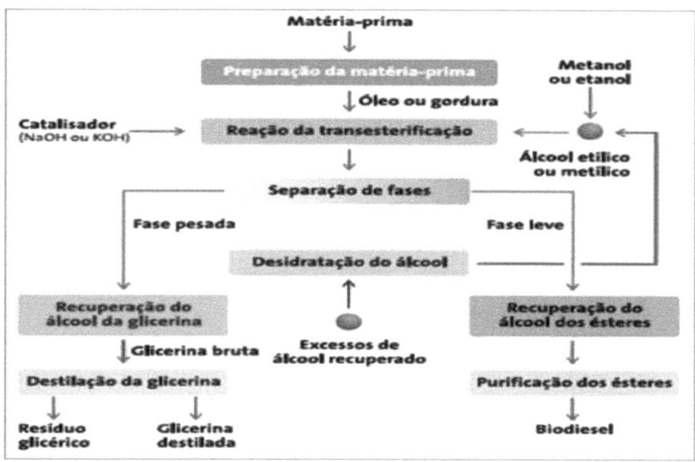

Figure 13: Obtaining biodiesel from transesterification
Source: Tn Sustainable

3.3. Use of castor oil cake and its applications.

3.3.1 Castor bean cake as animal feed

The residue from pressing the seeds after extracting the oil, known as the cake, could be used as animal feed and thus add commercial value to the castor bean (OLIVEIRA, 2007). The cake could be used after grinding to obtain bran, taking advantage of its high protein content in animal feed (RIBEIRO; ÁVILA, 2006). However, its use as animal feed has not been possible due to the presence of toxic and allergenic elements in its composition and the lack of technology, which makes it economically unviable at an industrial level to process it. On an experimental scale, however, the elimination of these toxic elements is easily achieved through thermal and chemical treatments. There are various methods to promote detoxification and deallergenisation of castor bean cake, using physical and chemical processes (RIBEIRO; ÁVILA, 2006). However, the production process was suspended due to the difficulty in controlling the efficiency of the detoxification process, which led to the release of batches of product that were still toxic and could cause the death of animals (SEVERINO, 2005). To date, the most effective way of eliminating ricin from castor beans is through genetic improvement, selecting seeds with a lower content of the toxin, through traditional breeding or even transgenic breeding.

Tratamento Físico

Agente	Concentração	Tempo	Remoção (%)
Embebição em água	10 L de água	3 h	65
		6 h	86
		12 h	84
Aquecimento com vapor	150 g de água	30 min	73
	(passagem de vapor)	60 min	85
Ebulição	10 L de água	30 min	90
	(fervura a 100ºC)	60 min	91
Autoclavagem (1,05 atm)	15 psi	30 min	65
		60 min	100
Forno de ar quente	100ºC	30 min	52
	120ºC	25 min	50

Tratamento Químico

Agente	Concentração	Tempo	Remoção (%)
NaOH	0,38 mol/L		86
	0,75 mol/L		91
NaCl	0,25 mol/L		82
	0,50 mol/L	8 horas	86
	1,00 mol/L		91
Ca(OH)2	10,00 g/kg		67
	20,00 g/kg	8 horas	68
	40,00 g/kg		100
Formaldeído	5,00 g/kg	7 dias	39
	10,00 g/kg		81
Amônia	7,50 g/kg	7 dias	51
	12,50 g/kg		59

Figure 14: Physical and chemical treatments to remove ricin

ANANDAN et al., 2005 adapted EMBRAPA

3.3.2. Castor bean cake as an organic fertiliser

Traditionally, this by-product has been used as an organic fertiliser due to its high nitrogen content. Its use as an agricultural input is appreciated by the

It provides organic matter to the soil and nutrients to crops, as well as its nematicidal effect, reducing the use of polluting agrochemical products. It also has a significant effect on the regeneration capacity of degraded soils, as it contains around 89% organic matter (SAVY, 2007). According to Lima et al (2006), castor bean cake used as an organic fertiliser on the CSRN 393 castor bean plant led to an increase in all growth characteristics, in proportion to the dose supplied. The leaf area, for example, increased from 342.45 to 1,024.85 cm2 as a result of the increase from 0.5 to 2.0 t/ha of castor bean cake. Total dry matter increased from 9.78 to 24.63 g between the highest and lowest doses. Castor bean cake had good characteristics for use as an organic fertiliser, mainly due to its high nitrogen content.

The mineralisation of castor bean cake is also much faster than that of cattle manure and sugarcane bagasse, according to results obtained by Severino (2004), which allows for a greater release of nutrients than other materials, but less availability when compared to chemical fertilisers. The faster decomposition of organic matter is probably due to the high levels of nitrogen, phosphorus and potassium present in the cake, as well as the optimum conditions for microbial activity: high humidity, good aeration and a temperature of around 28 °C (SEVERINO et al., 2004).

2.1.1. Castor bean cake as a nematicide

Castor bean cake has a nematicidal effect, making it a low-cost alternative to pesticides, which can cause damage to the operator and the environment. Studies suggest the use of castor bean cake as an alternative product for reducing nematode populations in the soil. According to Dutra et al (2006) in irrigated coffee trees, the application of castor bean cake to control nematodes can probably be attributed to the following effects: the action of the ricin-ricinin complex present in castor bean cake, which can be toxic to nematodes; the nutritional action promoted by castor bean cake; or the combined action of these effects. However, these results need to be complemented with specific studies to elucidate the biochemical, physiological and cytological mechanisms of the components of castor bean cake in reducing nematodes. The efficiency of the cake in combating the nematode Nacobbus aberrans was also verified in tomato plants, and the effect was attributed to the release of toxic compounds and lectin, ricin and ricin agglutinin (NAVARRO et al., 2002).

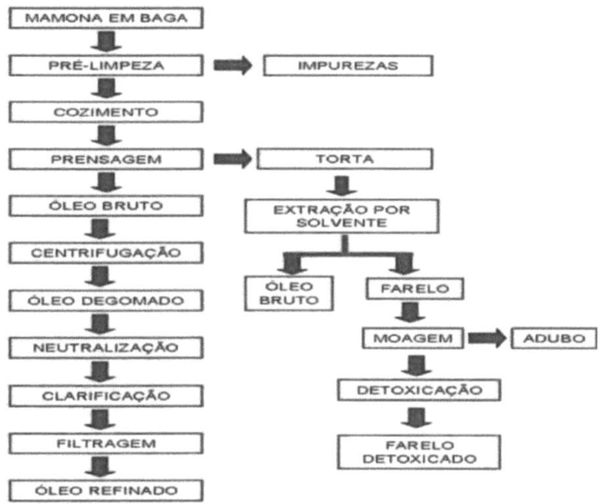

Figure 15: Flowchart of the castor oil, cake and fertiliser extraction process

Source: Embrapa

3.4. Castor oil as a biocidal agent

The use of synthetic insecticides has been the most widely adopted method for controlling insect pests. However, the various problems associated with the extensive use of these insecticides, such as poisoning and contamination, have led to the search for control methods that have less impact on the environment, such as the use of natural products extracted from plants (Perez & Iannacone, 2006 *apud* Bestete et al. 2011).

Plants are capable of synthesising substances that are harmful to herbivores, which can be extracted from their tissues and used as insecticides. The papaya tree is

one of these plants, whose oil extracted from the seed can cause up to 90 per cent mortality in larvae of *Diaphania nitidalis* Cramer (Lep.: Crambidae) after 72 hours of evaluation. It is known that natural insecticides degrade quickly under environmental conditions and that the association of entomopathogenic fungi with other control methods can provide synergistic action and faster pest control (Wiesbrook, 2004; Lima, 2009; *apud* Rondelli, et al. 2011).

Rondelli et al. (2011) reported the toxic effect of castor oil on *Plutellaxylostella* (cruciferous moth), by ingestion and contact, and attributed the mortality of *Plutella xylostella to the* presence of the toxin ricin, composed of two subunits, one of which is capable of binding to the enzyme alpha-amylase, thus preventing the digestion and absorption of starch, and the other subunit binds to ribosomes, inhibiting protein synthesis and causing the death of the moth. They also carried out experiments on the interaction between castor oil and *Beauveria bassiana* (a fungus capable of parasitising insects) and found that there was a significant effect of this association, showing synergism between *B. bassiana* and castor oil, indicating that the oil does not affect the pathogenicity of the fungus, and concluded that spraying this association between 2% castor oil and *B. bassiana with* 70 days of storage is efficient in controlling *P. xylostella.*

Bestete et al. (2011) observed that castor oil has a biocidal action when applied by contact and by ingestion, as it increases the mortality of *Helicoverpa zea (the tomato hornworm)*, although the effect of the oil by contact is greater than by ingestion when it comes to concentrations lower than 1.0%. They also state that castor oil at 3% is more lethal if ingested. Castor oil reduces the survival of *H. zea* larvae and does not harm the development of *Trichogramma pretiosum* (a species of small wasps that parasitise *H. zea* eggs), as long as it is applied after the parasitoid has been released, these two factors increase the possibility of exterminating the agricultural pest.

Messetti et al. (2010) showed that chemical derivatives of castor oil can be used to make a product that can be used as an alternative to control contamination by *Eleuconostoc mesenteroides* (a bacterium that secretes lactic acid and dextran, a viscous substance) in the sugar and alcohol industry. Castor oil proved to be effective in controlling the growth and development of the bacteria in the first six hours of application. The product acts efficiently against the contaminant, although there is less of a reduction in growth at pH 3.0, which is very common in the ethanol fermentation process.

Narciso (2014) carried out a treatment with water-based castor bean extract to analyse larval toxicity and the response was positive, as all the concentrations tested showed significant mortality against the larvae of A. aegypti, a well-known urban disease vector. Data from the study suggests that after 72 hours of exposure to the environment, the extract begins to decompose.

3.5. Castor oil-based polyurethane resins and their applications

After processing and synthesis, castor oil in the form of a polyol, in combination with a prepolymer, gives rise to a polyurethane, compatible with petrochemical polyurethanes, but with almost zero toxicity and low flammability. In foam form, it does not require external expanding agents to expand. Of plant origin, it is renewable and can be degraded in nature over time, compared to those of mineral origin (Lopes, 2009).

According to Marques and Martins (2009), polyurethane has been used in various areas of industry due to the possibility of obtaining this versatile polymer from four physical properties, making it possible to obtain infinite variations in characteristics by combining raw materials with different chemical properties, such as the choice between the various types of aromatic and aliphatic isocyanates found on the market.

Polyurethane derived from castor oil is a biodegradable polymer. Tests carried out on PU foams have shown that this material biodegrades well in the presence of microorganisms derived from biological agents that degrade fats.

Compared to petroleum polyurethanes, in cases where safety is a preponderant factor, such as in places where there is a risk of fire, its application is mandatory. As in the case of aeroplanes, this polymer is used both as a thermoacoustic insulator and as a flexible foam for use in the upholstery of seats, as it does not release toxic gases when burnt (Lopes, 2009).

According to Silva et al. (2015) epoxy resin is the main product used in the resinisation of ornamental stone, but it has negative characteristics such as high cost, inherent toxicity and not being biodegradable. Seeking to preserve the environment, vegetable oils have emerged as an alternative for obtaining resins with various applications.

Based on this principle, a polyurethane resin was synthesised from castor oil and according to the results obtained, the resin was able to replace epoxy resin, because the structure of the castor oil molecules and cardanol (a non-isoprenoid phenolic lipid used as a synthetic intermediate in the resin industry) has the ability to interact with the mineral albite, and demonstrates a higher interaction capacity than epoxy resin.

Epoxy resin is also one of the most widely used components in the manufacture of polymer composites. Seeking a biomass-based material for the manufacture of flooring and coverings, Oliveira (2011) analysed the feasibility of applying a composite based on castor oil polyurethane resin and Ubaçu (*Manicaria saccifera*) fibres. The composite was subjected to flexion, compression, impact and abrasion tests, among others, and the results showed that the material is suitable for the manufacture of flooring and coverings. As part of the same work, a floor was developed according to the Industrial Design methodology in the form of a "taco", which was selected for exhibition at the 24th Museu da Casa Brasileira Awards and in a Patent Register for the material by PUC Rio.

Beatrice and Vecchia (2011) report that during the construction of an experimental green roof, each platform underwent a surface waterproofing process using a castor oil-based polyurethane resin developed at the São Carlos Chemistry Institute (IQSC-USP).

The impregnation of the polyurethaneresin on the upper surface of the roofing slab made it possible to effectively plug the surface pores, permeabilising it and forming a highly compact film on the surface, promoting the encapsulation of the material. This experiment allowed the installation of the green roof to be successful, preventing water leakage, and is one of the demonstrations that prove that the resin obtained from castor oil can be used for surface waterproofing of roofs and buildings.

Figure 16 - Application of a waterproofing membrane based on castor oil (*Ricinus Communis*) on the experimental platforms.

Source: Beatrice and Vecchia (2011)

Cruz (2009) concluded that the use of castor oil polyurethane with an aggregate of thermosetting plastic waste proved to be a superior compound compared to materials currently used for thermal insulation of walls and roofs. The voids on the surface of castor oil polyurethane with a greater amount of waste trap air and increase the material's insulation capacity.

This material is also able to conduct less heat into rooms and offer greater thermal inertia than commercial thermal insulators (rock wool, glass wool and petroleum polyurethane), it has ecological advantages as it is made from a toxic waste product, the material is also renewable and biodegradable and has the ability to be moulded into any shape at a lower purchase cost compared to the thermal insulators mentioned above.

CHAPTER 4

Popular and medicinal uses

4.1. Castor oil in ethnobotany

Maciel et al. (2002) state that popular observations on the use and efficacy of medicinal plants make a significant contribution to publicising the therapeutic potential of plants, and also arouse the interest of researchers in areas such as botany, pharmacology, phytochemistry, among others. Enriching knowledge and intensifying the use of many plants.

There was a time when natural products began to have a mystical significance and industrialised societies began to consider the use of natural products as an option for uneducated people who did not trust their pharmacological value (Rates, 2001 *Apud* Bugno, 2006).

According to Coutinho et al. (2002) Knowledge about the therapeutic value of certain plant species has been passed down through the ages from generation to generation, forming, along with other practices, a medical system known as traditional.

The popular use of plants in traditional medicine is a valuable source for discovering new therapeutic agents, and maintains a prominent role in public health. However, popular knowledge must be associated with bioassays to confirm the efficacy and toxicity of these plants for therapeutic use (Bolde, 2004; Clardy & Walsh; 2004; Koehn & Carter, 2005 *Apud* Valderramas, 2006).

Medicinal plants can play an important role in health care and in some cases are safe and effective treatment alternatives. It is known that some plants can be indicated for the symptomatic relief of minor illnesses, and that medicines produced from fresh plants or

Dried plants or their derivatives can be commercialised in different pharmaceutical forms, such as syrups, solutions, tablets, ointments, gels and creams, with many routes of administration (BRASIL, 2008 *apud* Vargas and Arndt, 2014).

Souza et al. (2015) conducted 32 interviews and their reports revealed the use of 47 plant species in ethnoveterinary medicine in the region of Ivaiporã (PR), including *Ricinus communis,* which the interviewees recommended for cuts, wounds and breast infections. According to their findings, the interviewees used castor oil applied directly to the lesion for these ailments.

Ricinoleic acid, the base component of castor oil, has been shown to be an effective alternative in the treatment of wounds, capable of acting in the various stages of healing. It is also indicated as a chemical debridant (its function is to detach devitalised and fibrous tissues), a healing agent, keeps the wound bed moist, accelerates neo-angiogenesis and the granulation process and has a bactericidal action (Mandelbaum et al. 2003).

4.2. Uses of castor oil for medicinal purposes

This oil is also considered a natural polyester polyol that can be used to produce polyurethane (Alves, 2010). The polymer derived from castor oil is biocompatible and capable of promoting bone neoformation. It has many veterinary uses and in human medicine, there are many reports in which these polymers have been used to replace bones, for orthopaedic or dental purposes.

As it is a national product, low cost, non-toxic and biocompatible, its use is very useful in veterinary medicine, where costs often prevent an intervention (Laranjeira et al. 2004). In this case, making interventions more economically viable can reduce or prevent animal suffering and also create a new and attractive field of activity in the veterinary market.

Some experiments show that the polymer derived from castor beans has efficient applications in various areas. (Mastrantonio and Ramalho, 2003) carried out an experiment using polyurethane derived from castor bean resin to test its efficiency and biocompatibility with other substances in mouse connective tissue. The results of this study showed that this vegetable polyurethane is biocompatible, as only in the initial period (7 days) was there slight inflammation and also the presence of macrophages. The period of 7 days after surgery characterises the so-called synthesis period, with the deposition of hyaluronic acid and collagen produced by fibroblasts, and represents the replacement of injured structures. Similar results have been found in other studies, such as:

> "Castor bean polyurethane does not trigger infectious or rejection processes in the host tissue and is biocompatible, remaining biotolerant over time, without osseointegration" (Maria et al. 2003).

Peres et al. (2013) showed through experimental work the healing effect of using castor oil on equine wounds. A 17-year-old Quarter Horse had a wound that affected the skin, subcutaneous tissue and a slight area of muscle, initially 35 cm long and 19 cm wide. He reports that:

> "The measurements had an average reduction of 0.80 cm in width and 1.47 cm in length per week. The greatest healing speeds in width measurements were observed at the beginning of treatment and in length measurements at the end of treatment" (Peres et al. 2013).

Based on the results obtained, they state that the use of castor oil was effective in the treatment because it showed adequate healing, good product yield, easy application and no potential for bacterial resistance.

Valderramas (2006) revealed through an experimental study the anti-inflammatory effect of the polyurethane polymer obtained from *Ricinus communis* oil, using an inflammation-inducing agent on the inside of a mouse ear to produce oedema and erythema. Daily applications of the polymer showed that there were no side effects after 14 days of treatment, showing that the polymer lacks the toxicity common in *R. communis* oil. The study also showed that in both *in vivo* and *in vitro* models, *R. communis* has an efficient anti-inflammatory activity and capacity.

Vargas and Arndt (2014) tested solutions based on natural compounds with proven bacterial action as agents for disinfecting dentures, because they produce residues with a low environmental impact. They used a 2% solution of castor oil to immerse colourless acrylic resin and concluded that the roughness, micro-hardness and colour of the colourless acrylic resin were not altered after immersion, thus enabling efficient disinfection of dental prostheses.

(Pinelli et al. 2013 *apud* Vargas and Arndt, 2014) evaluated the effectiveness of 2% castor oil in the treatment of 30 elderly people with stomatitis, the performance of the oil was compared with the performance of nystatin and miconazole gel. It was concluded that treatment with castor oil was effective in reducing the clinical signs of stomatitis, and its effectiveness was similar to that of treatment with Miconazole. Although there was no simultaneous relationship between the reduction in the number of colonies and the improvement in clinical signs, castor oil has the ability to act against fungi and bacteria and thus probably reduce the virulence of the microflora.

CHAPTER 5

Ricinus in cosmetology and dermatology

Various plant extracts and compounds are used in cosmetology and dermatology, and *R. communis* has active principles that are also interesting for use in the production of various cosmetics.

Castor oil has an emollient, soothing, moisturising and protective effect on skin tissue, improving the appearance of rough and acne-prone skin. It also contains phytoestrogens and tocopherols that stimulate skin regeneration. Due to these properties, it is understood that the oil can be used in formulations aimed at cleansing and conditioning the skin (Aburjai and Natsheh 2008, *apud* Ruivo, 2012).

This oil is also widely used as a raw material for surfactants used in shampoos, which can be applied to fine hair where they exert a conditioning action. Undecylenic acid can be obtained from this oil, which because of its anti-mycotic action has been used as a preservative in cosmetics (Ruivo, 2012).

(Anthony, 2009 *apud Ruivo,* 2012) recommends using the oil for baby skin care and dry skin care, as it has a protective, softening action and a high degree of occlusivity.

Still on the cosmetic functions of castor oil, we can mention its use as an organic solvent to remove nail polish, applied in the form of a 30 per cent solution it replenishes the lipids of the nail, helping with its maintenance and aesthetics (Batata, 2002 *apud* Ruivo, 2012).

These are just some of the cosmetic uses of *Ricinus communis.* It is worth remembering the importance of studies by biologists and cosmetologists in order to increase the safety of using this plant species in formulations without causing harm to those exposed to the product. Studies on this species can cater for the public who prefer products of plant origin.

CHAPTER 6

Industrial use of the *R.communis* leaf

Industry has needs that must be met in order for the production process to run smoothly, and by meeting these needs we can reduce costs and increase profits. Green alternatives to replace products that are constantly needed are becoming more popular because they represent greener and healthier options. Having demonstrated so many applications for R. *communis, it* is easy to conclude that it is one of the species that can contribute to this important objective.

The *R. communis* leaf extract inhibits steel corrosion by 84% when in aqueous solution at a concentration of 300ppm in NaCl. Polarisation measurements indicated that the extract acts as an anodic inhibitor and electrochemical impedance measurements showed that the formation of an organic iron-compound complex on the steel surface is responsible for reducing corrosion in neutral solution (Sathiyanathan et al. 2005 *apud* De Araujo Neto et al. 2013).

Based on its inhibition percentage, we can say that this extract is a good green inhibitor. Its use has a good ecological impact, as it is a biodegradable product and the use of the plant's leaf makes its production more economically attractive, as the producer can make a profit from various parts and by-products of the plant, making it possible to generate good development in producing regions. The possibility of producing a variety of products with added value from a single species makes castor bean cultivation very attractive economically.

CHAPTER 7

Final considerations

This work brings together information on how to take advantage of the versatility of castor oil (*Ricinus communis* L.) in traditional medicine, modern medicine, biotechnology and industry. We recognise, however, that due to the limitations of the study, we have only been able to cover a few of the applications of castor oil, as it is still difficult to find up-to-date literature describing the applications of castor oil in detail, especially in Portuguese. We did not cover topics such as its use in fibre optics and its use as antifreeze in aircraft, due to the lack of detailed and reliable sources.

In our bibliographical survey, we observed a selection of experience reports and experimental studies that could contribute to future studies on castor oil, since such a versatile raw material still needs to be studied and subsequently used as a way of solving everyday problems in our society.

We look at the stages in the process of producing biodiesel from castor beans, highlighting the negative and positive characteristics of using castor beans as a raw material and some of the social aspects involved in the production and harvesting of castor beans.

We have also made space to mention some of the chemical characteristics of castor oil that make it possible to use it in various areas. These applications are described in such a way as to relate different studies on the same application and compare the results obtained, so that the reader can have a view from different angles.

Another highly relevant topic that has been addressed is the toxicity of castor bean, which has applications in pest and vector control, among others. Toxicity, which for a long time was seen as a negative and dangerous characteristic of the species, has been addressed in more recent research as a more natural alternative to solve problems in urban and rural environments.

Some little-known applications of castor bean also received attention, such as: use as a thermal insulator, use in the treatment and healing of injuries, bactericidal action, antifungal action. We would like to highlight applications in various

formulations and showed that *Ricinus communis* can be a plant-based alternative for improving products in various market segments, as each part of the plant can be used in different ways, making it economically attractive and increasing the interest of farmers, fostering new market segments and expanding the development of regions able to produce the plant species and possibly various products derived from it and with added value.

CHAPTER 8

Bibliographical references

* ALMEIDA, Bruno Melegari De Souza; DOS SANTOS, Robson Luiz Costa; JÚNIOR, Antonio Fluminhan. GENETIC IMPROVEMENT OF MAMON (Ricinus communis L.) WITH A VIEW TO PRODUCING VARIETIES ADAPTED TO MECHANISED CULTIVATION. **Electronic Journal Fórum Ambiental da Alta Paulista**, v. 11, n. 2, 2015.

* ALVES, William Ferreira. Preparation and mechanical, thermal and electrical characterisation of mixtures of polyurethane derived from castor oil and poly (o-methoxyaniline) for evaluation in application as sensors for electronic tongues. 2010.

* ANANDAN, A.; KUMAR, G. K. A.; GHOSH, J.; RAMACHANDRA, K. S. **Effect of different physical and chemical treatments on detoxification of ricin in castor cake.** Animal Feed Science and Technology, Amsterdam, v.120, n. 12, p.159-168, 2005.

* BEATRICE, Caio Cury; VECCHIA, Francisco. EVALUATION OF THE POTENTIAL USE OF THREE PLANT SPECIES AS LIGHTWEIGHT ROOF COVERINGS FOR BUILDINGS. **Revista de Ciências Ambientais**, v. 5, n. 1, p. 5-24, 2011.

* BELTRÃO, NE de M.; DE OLIVEIRA, M. I. P. Oilseeds and their oils: advantages and disadvantages for biodiesel production. **Embrapa Algodão- Documentos (INFOTECA-E)**, 2008.

* BESTETE, Luziani Rezende, Pratissoli, D., Queiroz, V. T. D., Celestino, F. N., & Machado, L. C. Toxicity of Castor bean oil on Helicoverpa zea and Trichogramma pretiosum. **Pesquisa Agropecuária Brasileira**, v. 46, n. 8, p. 791-797, 2011.

* BUGNO, Adriana. **Plant drugs: evaluation of microbial contamination and research into aflatoxins, ochratoxin A and citrinin.** 2006. Doctoral thesis.

University of São Paulo.

• COUTINHO, D. F.; TRAVASSOS, L. M. A.; DO AMARAL, F. M. M. **Estudo etnobotânico de plantas medicinais utilizadas em comunidades indígenas no estado do maranhão - BRASIL**. Visão Acadêmica, Curitiba, v. 3, n. 1, p. 1-12, Jan.-Jun./2002.

• CRUZ, Michelle Paiva. **Application of industrial waste for thermal insulation: a proposal for using castor bean polyurethane with thermosetting plastic waste aggregate**. 2009. Master's dissertation. Federal University of Rio Grande do Norte.

• DA SILVA, Vanessa Quitete Ribeiro. **Alternative oilseed species for biodiesel production.** Available at:< https://www.embrapa.br/documents/1354377/1753093/Especies-oleaginosas-alternativas-producao-biodiesel-Vanessa-Quitete.pdf/10bff51f-a328-4cdd- b4f2-dbeab26435a4?version=1.0>. Accessed on: 04 August 2017.

• DE ARAÚJO NETO, José Amilcar Mendes. Caatinga plant extracts as corrosion inhibitors.2013.

• EMBRAPA. **Production Chain**: Castor Oil and Biodiesel. Available at: <Brasilhttp://www.cnpa.embrapa.br/produtos/mamona/cadeia_produtiva_biodi esel.html>. Accessed on 21 November 2017.

• FERREIRA, Igor Bezerra; MARTINS, Janaína Jenuário; DUARTE, Fábio Teixeira. Ricinus communis: THE MAMON AND ITS SYSTEMS BIOLOGY. In: **IX IFRN Scientific Initiation Congress**. 2013.

• GÓES, Paulo Sérgio de Assis. **O Papel da Petrobras na produção de Biodiesel: Perspectivas de produção e distribuição do biodiesel de mamona-**. Department of Environmental Energy - University of Bahia - Salvador, 2006.

• HOFFMANN, L. V. et al. Ricin: an impasse in the utilisation of castor bean cake and its applications. **Embrapa Cotton-Documents (INFOTECA- E)**, 2007.

* LARANJEIRA, Maria Gisela, Rezende, C. M. F., Sá, M. J. C., & Silva, C. M. Implants of vegetable polyurethane resin (Ricinus communis) in linear traction, fixation and vertebral fusion in the dog: experimental study. **Arquivo Brasileiro de Medicina Veterinária e Zootecnia**, p. 602-609, 2004.

* LIMA, R. L. S.; SEVERINO, L. S.; ALBUQUERQUE, R. C.; BELTRÃO, N. E. M. . Evaluation of castor bean husk and cake as an organic fertiliser. n: CONGRESSO BRASILEIRO DE MAMONA, 2., 2006, Aracaju, SE.

* LOPES, Edmar Maria Lima et al. **Thermal performance of castor oil-based polyurethane foam used in building components (under-coverings)**: a study in Ilha Solteira, SP. 2009.

* MACIEL, Maria Aparecida M. et al. **Medicinal plants: the need for multidisciplinary studies**. Química nova, v. 25, n. 3, p. 429-438, 2002.

* MANDELBAUM, Samuel Henrique; DI SANTIS, Érico Pampado; MANDELBAUM, Maria Helena Sant'Ana. Cicatrization: current concepts and auxiliary resources-Part II Cicatrization: current concepts and auxiliary resources-Part II. **An Bras Dermatol**, v. 78, n. 5, p. 525-542, 2003.

* MARIA, Patricia Popak; PADILHA FILHO, João Guilherme; CASTRO, Márcio Botelho. Macroscopic and histological analysis of the use of polyurethane derived from castor oil (Ricinus communis) applied to the tibia of growing dogs. **Acta Cirúrgica Brasileira**, p. 332-336, 2003.

* MARQUES, Bruno R.; MARTINS, L. J. R. Polyurethane Derived from Castor Oil: From the Environment to Biocompatibility. **Lins, SP,** 2009.

- MASTRANTONIO, Simone Di Salvo; RAMALHO, Lizeti Toledo de Oliveira. Response of mice connective tissue to castor oil vegetable polyurethane. **Rev. odontol. UNESP**, v. 32, n. 1, p. 31-37, 2003.

- MELO, Walber Carvalho, Silva, D. B. D., Júnior, N. P., Anna, L. M. M. S., & Santos, A. D. Production of ethanol from castor bean cake (Ricinus communis L.) and evaluation of the lethality of hydrolysed cake to mice. **Química Nova**, v. 31, n. 5, p. 1104-1106, 2008.

- MESSETTI, M. A., Santos, A. M., Angelis, D. F., Chierice, G. O., & Claro Neto, S. Study of Ricinus communis L.(castor bean) oil derivative as a biocidal agent and viscosity reducer produced by Leuconostoc mesenteroides in sugar-alcohol industries. **Arq. Inst. Biol., São Paulo**, v. 77, n. 2, p. 301-308, 2010.

- MILANI, Máira. **Castor bean knowledge tree.** Available at:<http://www.agencia.cnptia.embrapa.br/gestor/mamona/arvore/CONT000h 4rb0y9002wx7ha0awymty4m52beo.html>. Accessed on: 02 August 2017.

- MIRANDA, Júlio César de Carvalho et al. Creation of a database, simulation and energy analysis of the biodiesel production process from soya, castor beans and jatropha. 2011.

- MINISTRY OF MINES AND ENERGY. **National Programme for the Production and Use of Biodiesel** Available at: <http://www.mme.gov.br/programas/biodiesel/menu/biodiesel/perguntas.html> . Accessed on: 02 August 2017.

- MORO, Edemar. Nitrogen fertiliser management in low-growing castor bean hybrids grown in the harvest and off-season in a no-till system. 2008.

- NARCISO, JULIANA OLIVEIRA ABREU. Castor oil cake (Ricinus communis L); as a biolarvicide against Aedes aegypti L.(Diptera: Culicidae).2014.

- NAVARRO, F. F.; VERA, I. C. D.; MEJIA, E. Z.; GARCIA, P. S. Aplicación de enmiendas orgánicas para elmanejo de nacobbusaberransen tomate. Nematropica, Bradenton, v. 32, n. 2, 2002.

- OLIVEIRA, L. B; COSTA, A. O. Biodiesel: an experiment in sustainable development. Available at: Accessed on: 9 Jan. 2007.

- Oliveira, Ana Karla Freire de; D'ALMEIDA, Prof José Roberto Moraes. **Study of the technical feasibility of using the polyurethane composite of castor bean resin and ubuçu fibre in the manufacture of floor and wall coverings**. 2011. Doctoral thesis. PUC-Rio.

- OLIVEIRA FILHO, Antonio Francelino de. **Castor bean x food crop consortium in different spatial arrangements**. 2013. Doctoral thesis.

- OLIVEIRA, Tiara Gomes de. **Development of biodegradable films based on castor bean cake protein (Ricinus communis L.) modified with glyoxal and reinforced with cellulose fibres**. 2013. PhD Thesis. University of São Paulo.

- PERES, Anelise Ribeiro, A. R., de Melo Júnior, J., Ifran, A. M., Moda, T. F., & Casas, V. F. REPORT OF THE USE OF RICIN OIL AS A WOUND CARE IN EQUINES.2013.

- RIBEIRO, N. M.; AVILA, F. D. F. Methods for detoxifying oilseed cake. In: CONGRESS OF THE BRAZILIAN BIODIESEL TECHNOLOGY NETWORK, 2007, Brasília, DF. Proceedings... Brasília, DF: MCT/ABIPTI, 2007. , v. 1, p.34-37, 2006.

- RIZZI, B. SILVA; SILVA, G. A. J.; MAIOR, Thales Souto. Castor oil as a biofuel. **Journal of the Petrobras University and IF Fluminense Project**, v. 1, p. 317-320, 2010.

- RONDELLI, Vando Miossi, Rondelli, V. M., Pratissoli, D., Polanczyk, R. A., Marques, E.

J., Sturm, G. M., & Tiburcio, M. O. Association of castor oil with Beauveria bassiana in the control of the cruciferous moth. **Pesquisa Agropecuária Brasileira**, v. 46, n. 2, p. 212214, 2011.

- RUIVO, Joana Sofia Pais. Phytocosmetics: application of plant extracts in cosmetics and dermatology. **Porto: Fernando Pessoa University**, 2012.

- SAVY, F. A. Castor bean (Ricinus communis) - development of production technology, 2007 (PÉTER MURÁNYI AWARD).

- SEVERINO, L. S. What we know about castor bean cake. Campina Grande: Embrapa Algodão, 2005 (Embrapa Algodão. Documentos, 136).

- SCHNEIDER, Rosana de Cassia de Souza. Extraction, characterisation and transformation of castor oil. p. 28, 2003.

- SEVERINO, L. S.; COSTA, F. X.; BELTRÃO, N. E. M.; LUCENA M. A.; GUIMARÃES, M. M. B. Mineralisation of castor bean cake, cattle manure and sugarcane bagasse estimated by microbial respiration. Revista de Biologia e Ciências da Terra, Campina Grande, v. 5, n. 1,2004.

- SILVA, Fernanda Barbosa da; CORREIA, Júlio Cesar Guedes; CARAUTA, Alexandre Nelson Martiniano. **Evaluation of the efficiency of resins from renewable sources in the ornamental stone resin process using molecular modelling**. 2015.

- SILVA DE LIMA, Rosiane de Lourdes et al. Castor bean husk and cake evaluated in pots as organic fertilisers. **Revista Caatinga**, v. 21, n. 5, 2008.

- SOUZA, Aline, JÚNIOR, J. B. S., TEIXEIRA, E. J. R., LEN, L. S., MOURO, G. F., & MONTEIRO, V. Ethnoveterinary in the Region of Ivaiporã-PR: Popular Knowledge Contributing to the Construction of More Sustainable Animal Production Systems. **Cadernos de Agroecologia**, v. 9, n. 4, 2015.

- VALDERRAMAS, Adriana Cristina. Study of the anti-inflammatory activity of ricinus communis (Euphorbiacea). p.08-30, 2006.

* VARGAS, Clarissa Machado; ARNDT, Paula Borges. Effect of immersion in solutions of rosemary oil, castor oil and glycolic extract of propolis on the properties of a colourless acrylic resin: longitudinal study. 2014.

I want morebooks!

Buy your books fast and straightforward online - at one of world's fastest growing online book stores! Environmentally sound due to Print-on-Demand technologies.

Buy your books online at
www.morebooks.shop

Kaufen Sie Ihre Bücher schnell und unkompliziert online – auf einer der am schnellsten wachsenden Buchhandelsplattformen weltweit! Dank Print-On-Demand umwelt- und ressourcenschonend produziert.

Bücher schneller online kaufen
www.morebooks.shop

Printed by Books on Demand GmbH, Norderstedt / Germany